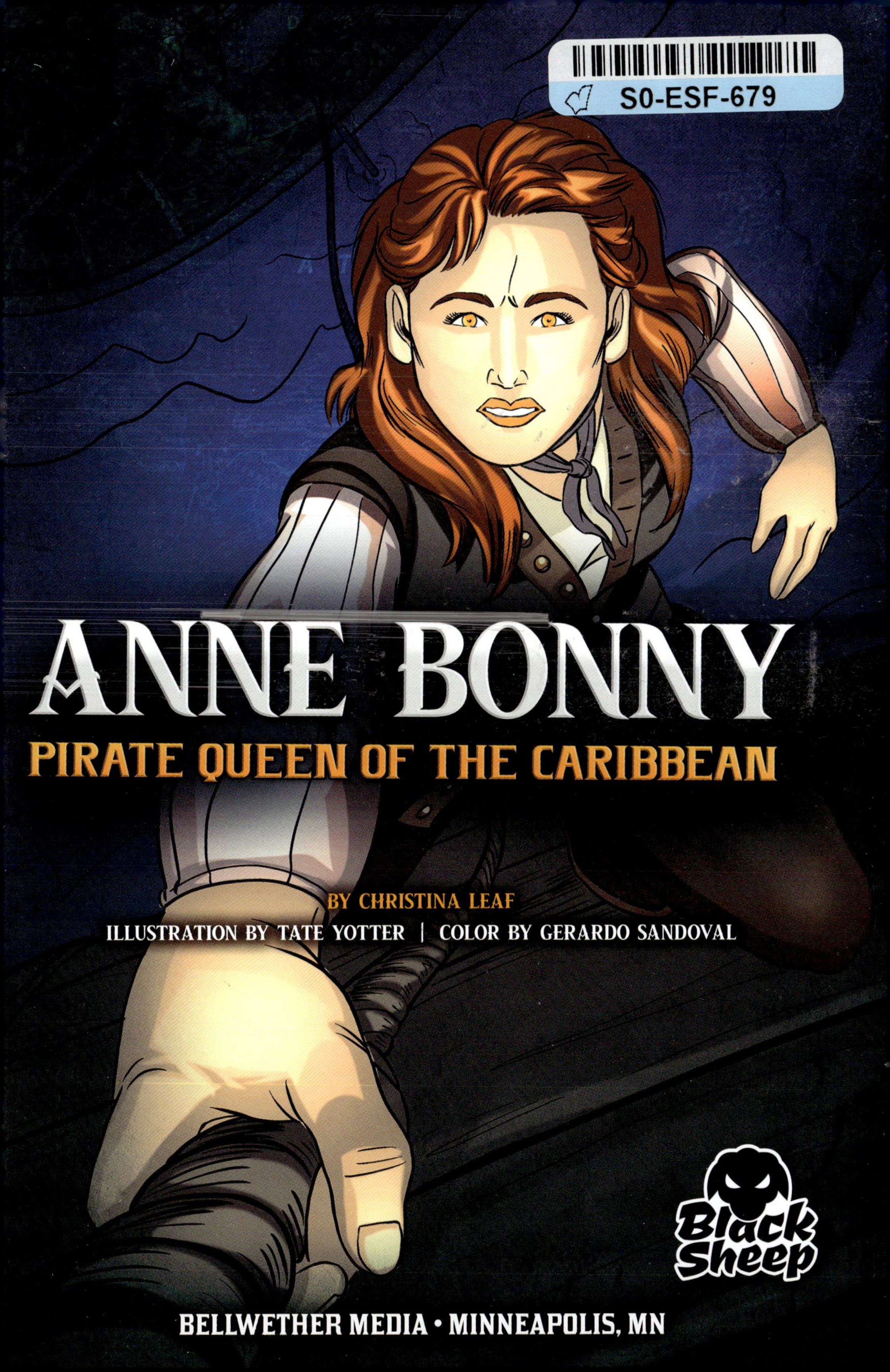

ANNE BONNY

PIRATE QUEEN OF THE CARIBBEAN

BY CHRISTINA LEAF

ILLUSTRATION BY TATE YOTTER | COLOR BY GERARDO SANDOVAL

Black Sheep

BELLWETHER MEDIA • MINNEAPOLIS, MN

Stray from regular reads with Black Sheep books. Feel a rush with every read!

This edition first published in 2021 by Bellwether Media, Inc.

No part of this publication may be reproduced in whole or in part without written permission of the publisher. For information regarding permission, write to Bellwether Media, Inc., Attention: Permissions Department, 6012 Blue Circle Drive, Minnetonka, MN 55343.

Library of Congress Cataloging-in-Publication Data

Names: Leaf, Christina, author.
Title: Anne Bonny : pirate queen of the Caribbean / by Christina Leaf.
Description: Minneapolis, MN : Bellwether Media, 2021. | Series: Black sheep. Pirate tales | Includes bibliographical references and index. | Audience: Ages 7-13 | Audience: Grades 4-6 | Summary: "Exciting illustrations follow events in the life of Anne Bonny. The combination of brightly colored panels and leveled text is intended for students in grades 3 through 8"– Provided by publisher.
Identifiers: LCCN 2020017785 (print) | LCCN 2020017786 (ebook) | ISBN 9781644873007 (library binding) | ISBN 9781681038391 (paperback) | ISBN 9781681037578 (ebook)
Subjects: LCSH: Bonny, Anne, 1700–Juvenile literature. | Women pirates–Caribbean Sea–Biography–Juvenile literature.
Classification: LCC G537.B67 L43 2021 (print) | LCC G537.B67 (ebook) | DDC 910.4/5 [B]–dc23
LC record available at https://lccn.loc.gov/2020017785
LC ebook record available at https://lccn.loc.gov/2020017786

Text copyright © 2021 by Bellwether Media, Inc. BLACK SHEEP and associated logos are trademarks and/or registered trademarks of Bellwether Media, Inc.

Editor: Betsy Rathburn   Designer: Andrea Schneider

Printed in the United States of America, North Mankato, MN.

TABLE OF CONTENTS

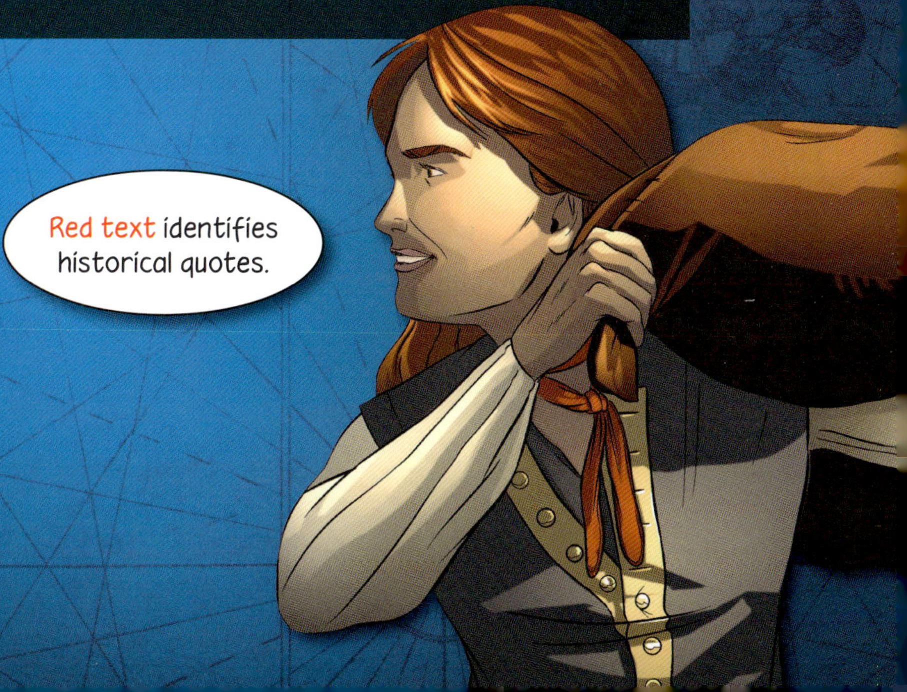

Red text identifies historical quotes.

A Life of Rebellion

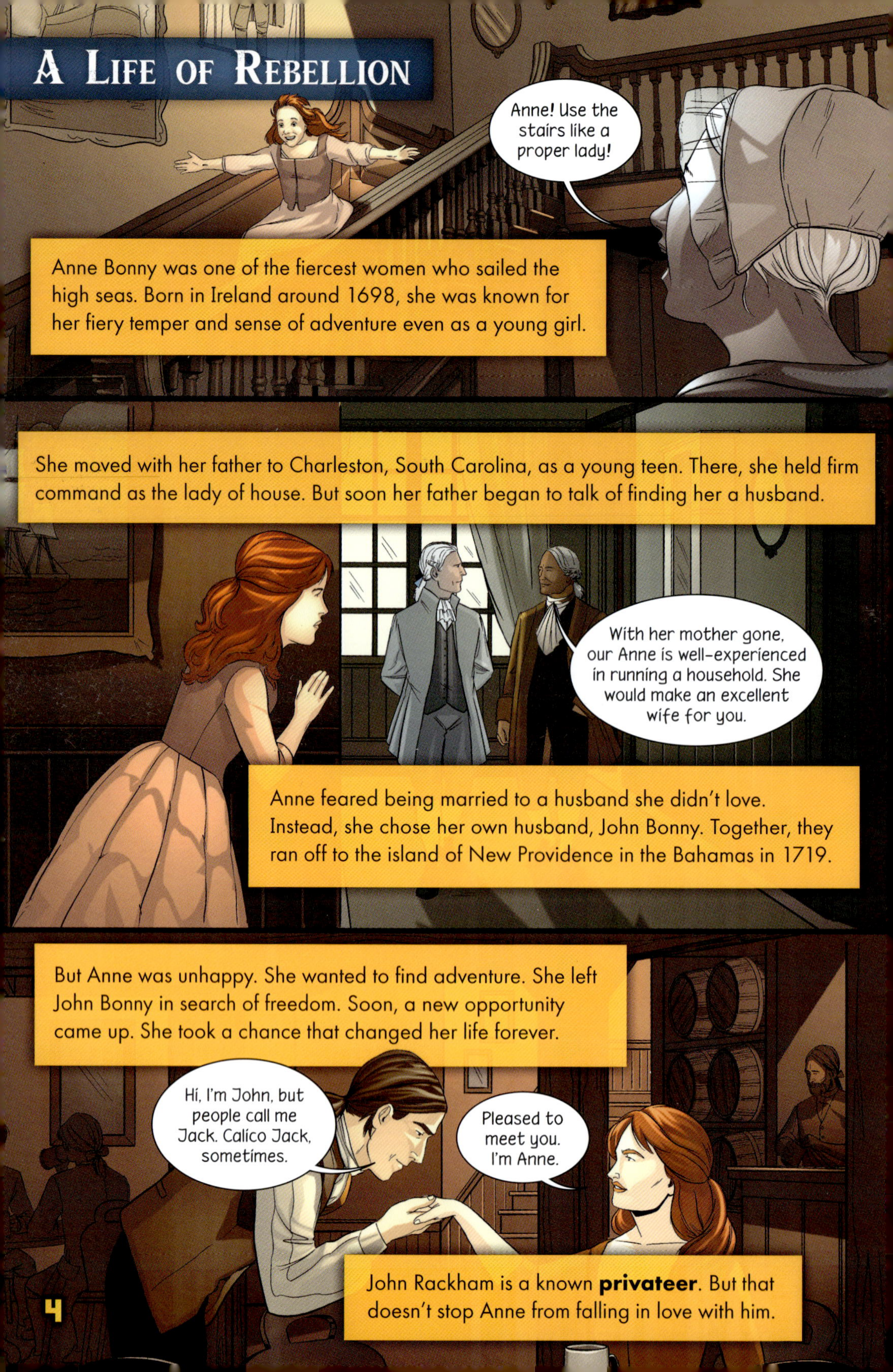
Anne! Use the stairs like a proper lady!
Anne Bonny was one of the fiercest women who sailed the high seas. Born in Ireland around 1698, she was known for her fiery temper and sense of adventure even as a young girl.
She moved with her father to Charleston, South Carolina, as a young teen. There, she held firm command as the lady of house. But soon her father began to talk of finding her a husband.
With her mother gone, our Anne is well-experienced in running a household. She would make an excellent wife for you.
Anne feared being married to a husband she didn't love. Instead, she chose her own husband, John Bonny. Together, they ran off to the island of New Providence in the Bahamas in 1719.
But Anne was unhappy. She wanted to find adventure. She left John Bonny in search of freedom. Soon, a new opportunity came up. She took a chance that changed her life forever.
Hi, I'm John, but people call me Jack. Calico Jack, sometimes.
Pleased to meet you. I'm Anne.
John Rackham is a known **privateer**. But that doesn't stop Anne from falling in love with him.

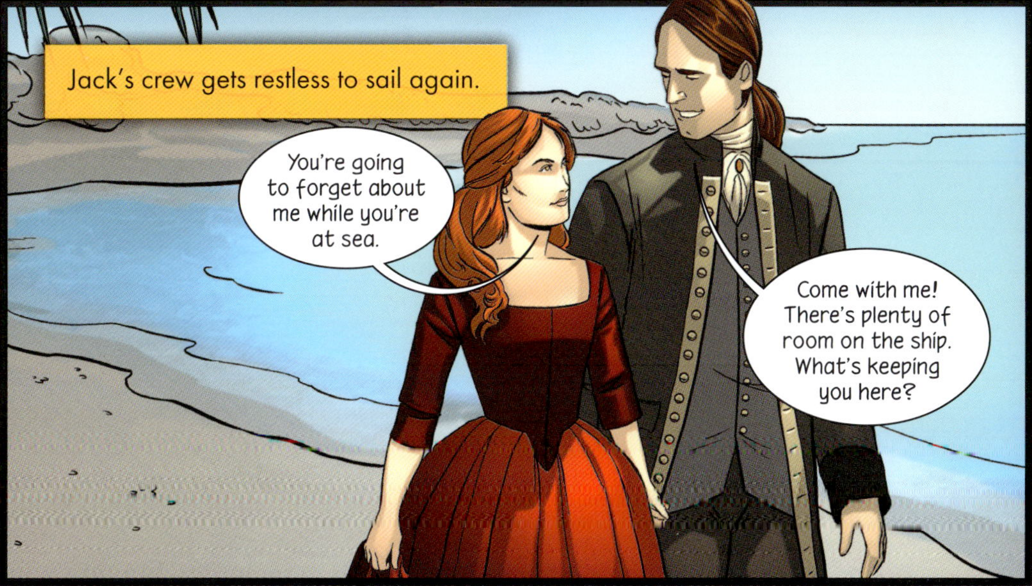
Jack's crew gets restless to sail again.
You're going to forget about me while you're at sea.
Come with me! There's plenty of room on the ship. What's keeping you here?

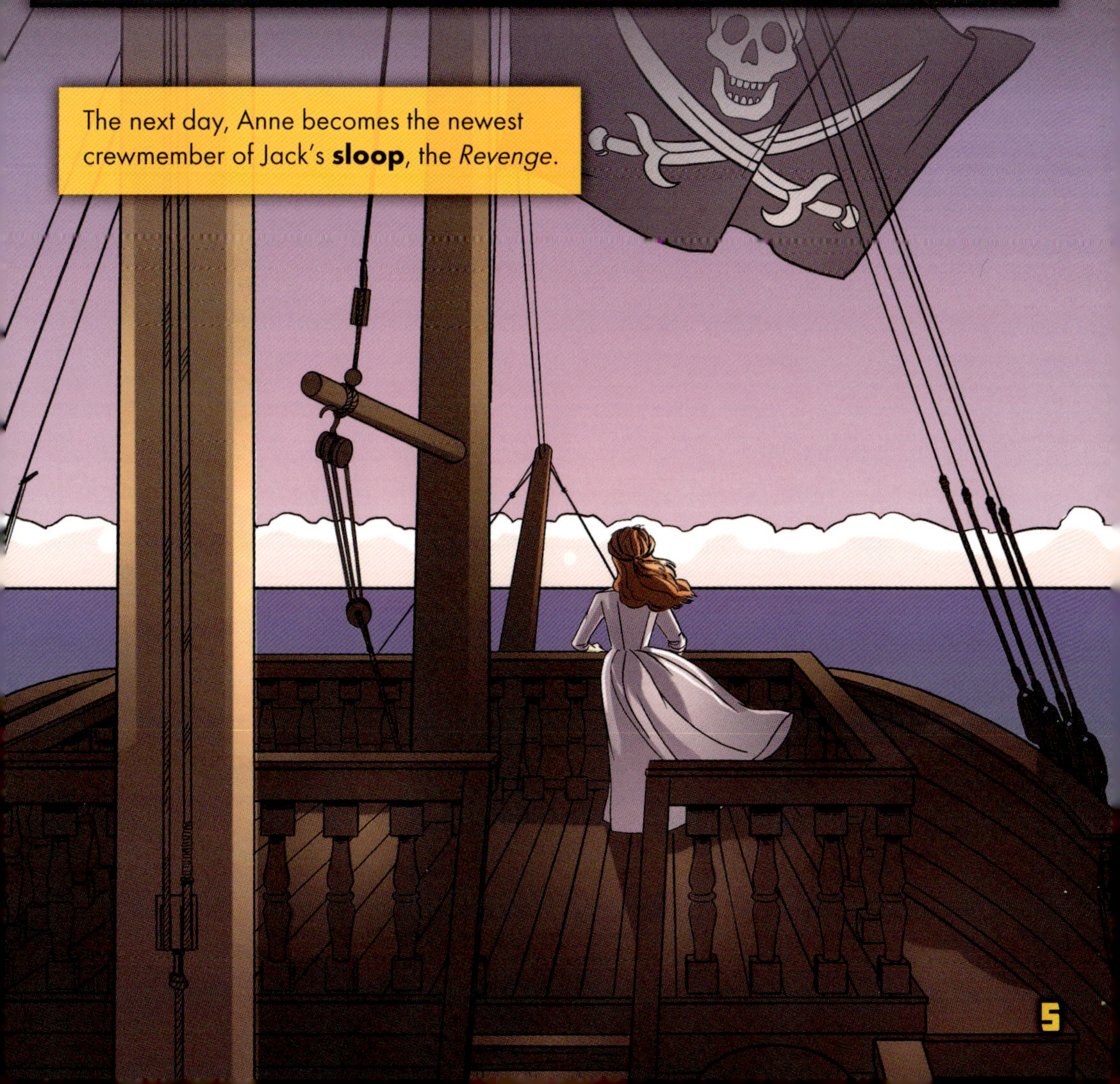
The next day, Anne becomes the newest crewmember of Jack's **sloop**, the *Revenge*.

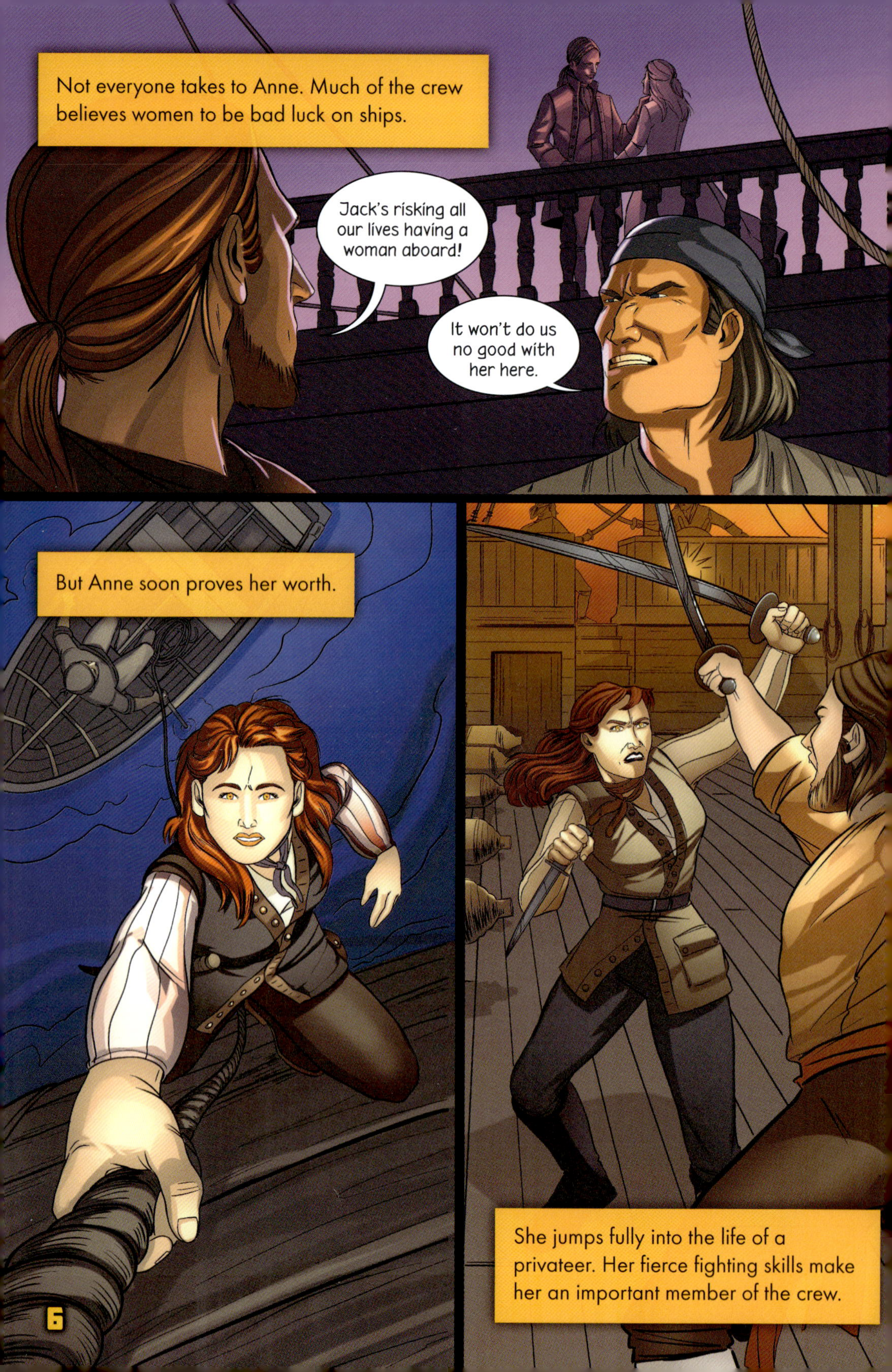
Not everyone takes to Anne. Much of the crew believes women to be bad luck on ships.
Jack's risking all our lives having a woman aboard!
It won't do us no good with her here.
But Anne soon proves her worth.
She jumps fully into the life of a privateer. Her fierce fighting skills make her an important member of the crew.

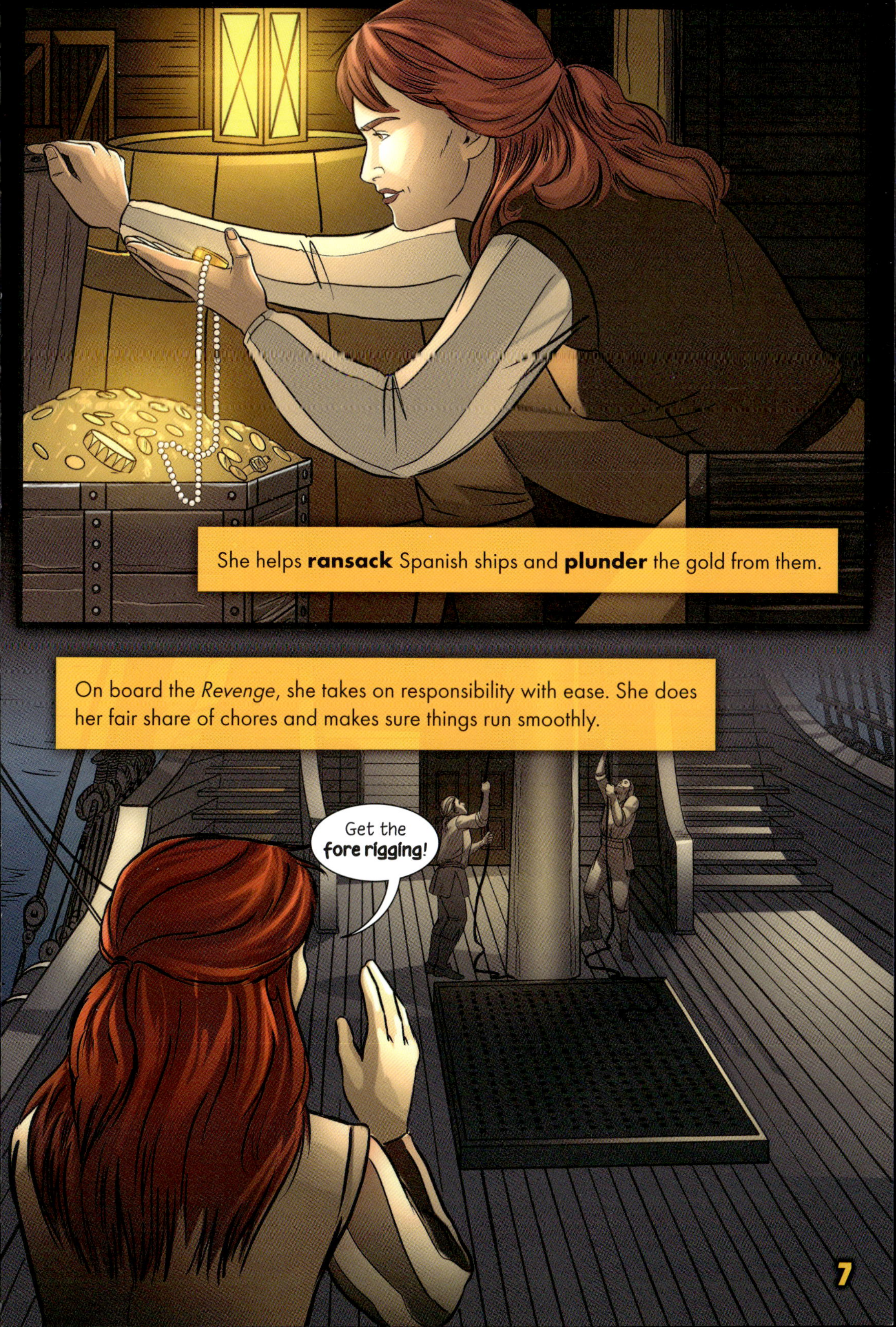
She helps **ransack** Spanish ships and **plunder** the gold from them.
On board the *Revenge*, she takes on responsibility with ease. She does her fair share of chores and makes sure things run smoothly.
Get the **fore rigging!**

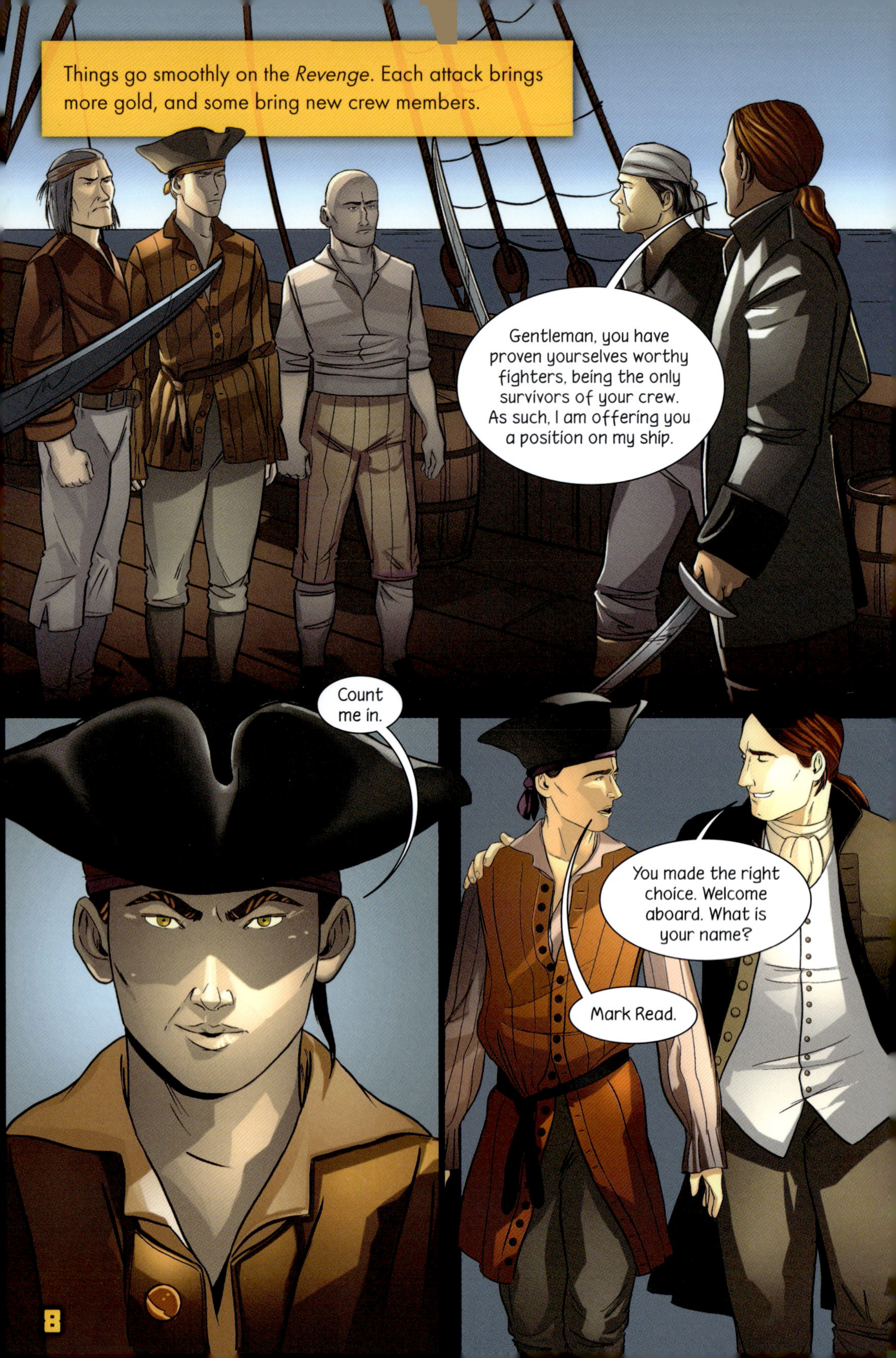
Things go smoothly on the *Revenge*. Each attack brings more gold, and some bring new crew members.
Gentleman, you have proven yourselves worthy fighters, being the only survivors of your crew. As such, I am offering you a position on my ship.
Count me in.
You made the right choice. Welcome aboard. What is your name?
Mark Read.

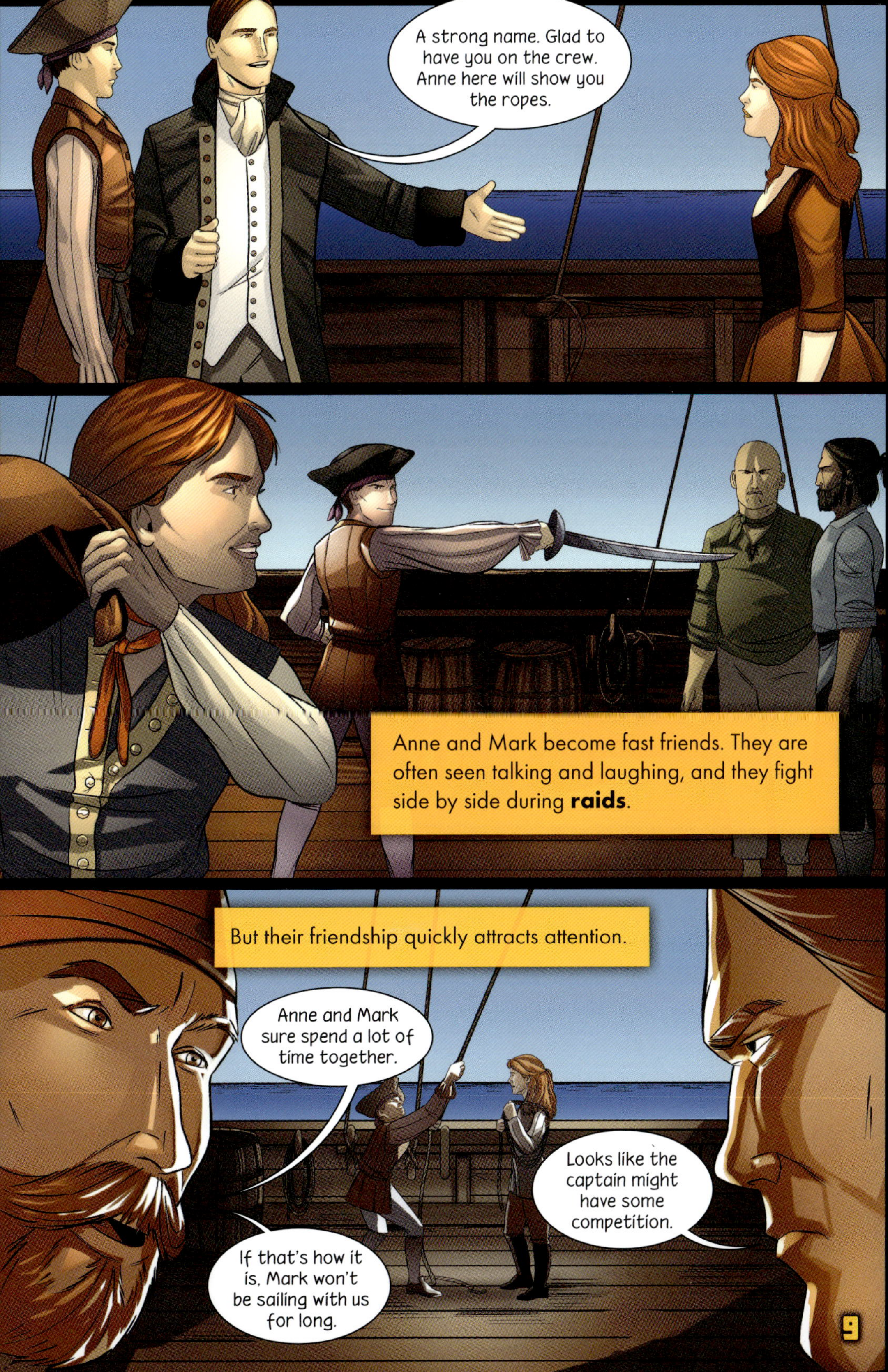
A strong name. Glad to have you on the crew. Anne here will show you the ropes.
Anne and Mark become fast friends. They are often seen talking and laughing, and they fight side by side during **raids**.
But their friendship quickly attracts attention.
Anne and Mark sure spend a lot of time together.
Looks like the captain might have some competition.
If that's how it is, Mark won't be sailing with us for long.

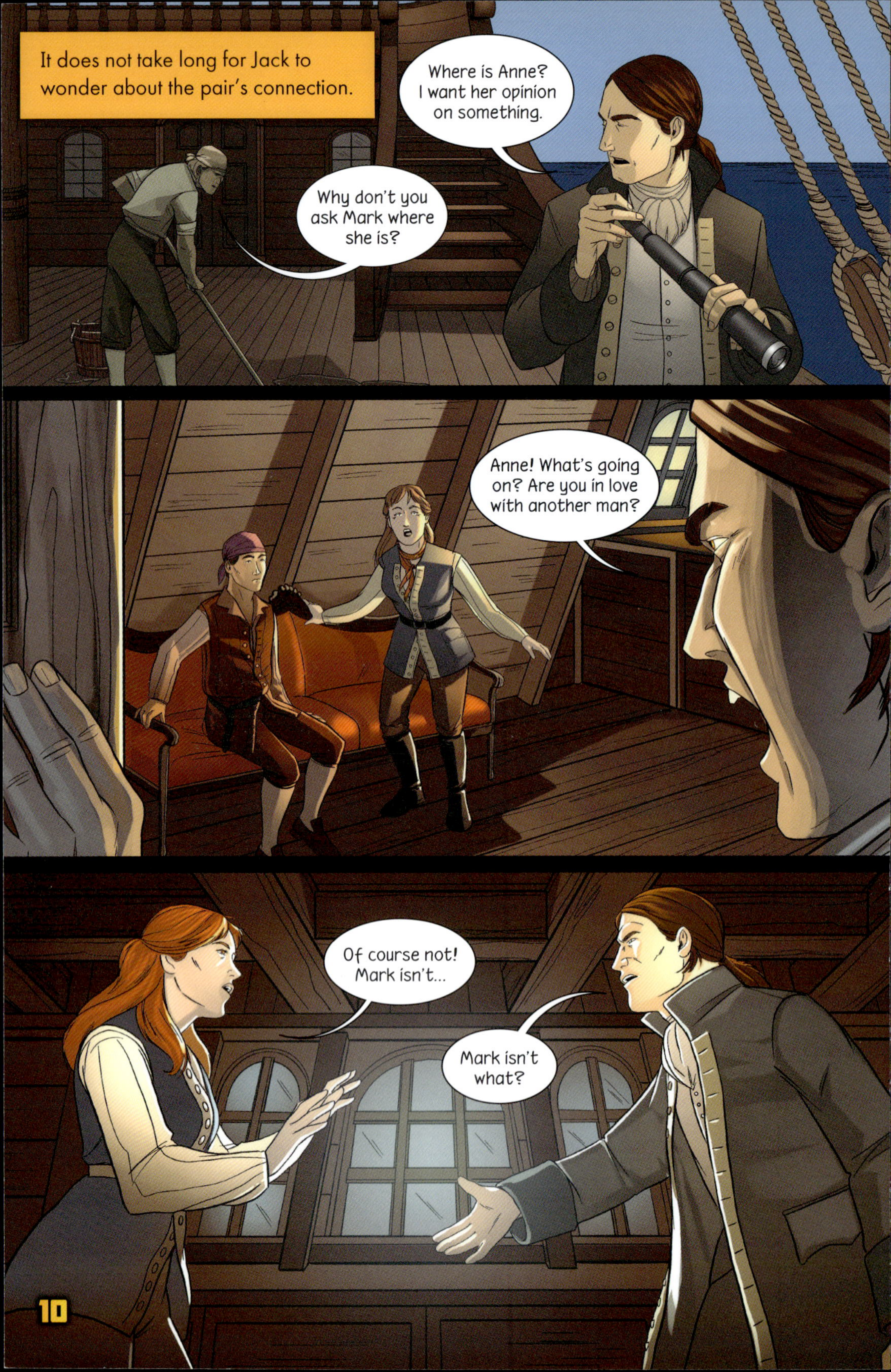
It does not take long for Jack to wonder about the pair's connection.
Where is Anne? I want her opinion on something.
Why don't you ask Mark where she is?
Anne! What's going on? Are you in love with another man?
Of course not! Mark isn't...
Mark isn't what?

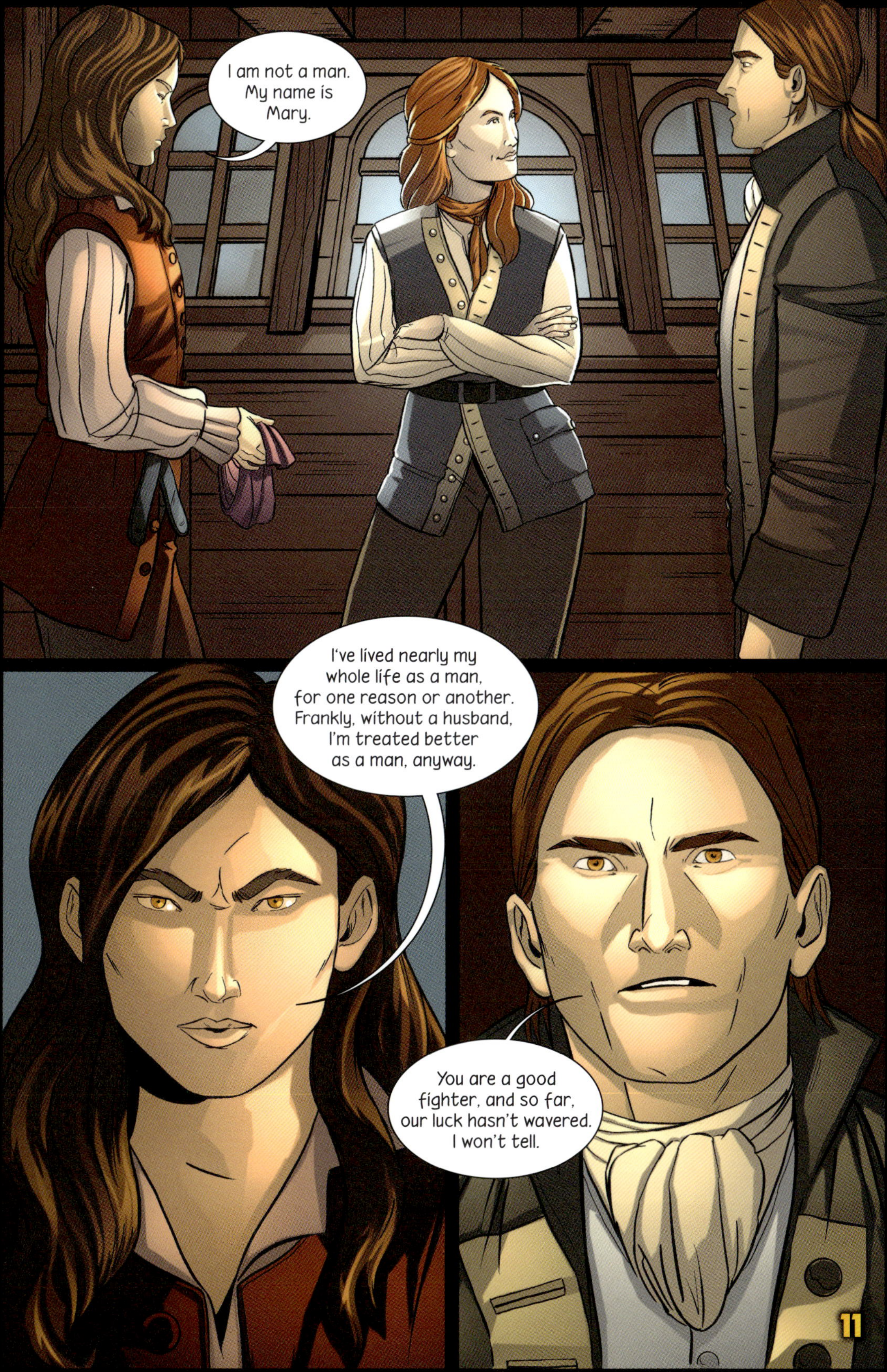
I am not a man. My name is Mary.
I've lived nearly my whole life as a man, for one reason or another. Frankly, without a husband, I'm treated better as a man, anyway.
You are a good fighter, and so far, our luck hasn't wavered. I won't tell.

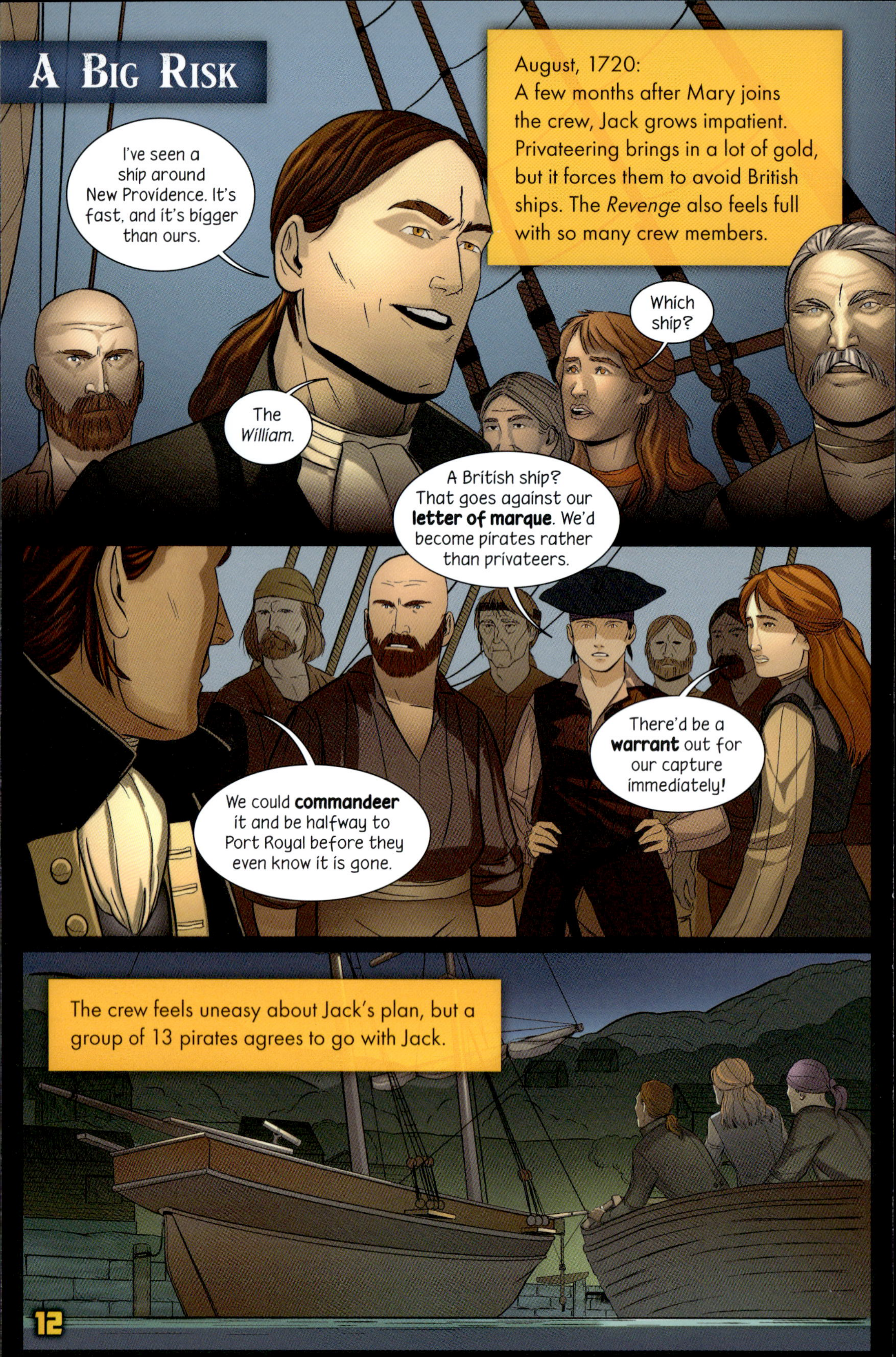
A Big Risk
August, 1720:
A few months after Mary joins the crew, Jack grows impatient. Privateering brings in a lot of gold, but it forces them to avoid British ships. The *Revenge* also feels full with so many crew members.
I've seen a ship around New Providence. It's fast, and it's bigger than ours.
Which ship?
The *William.*
A British ship? That goes against our **letter of marque**. We'd become pirates rather than privateers.
We could **commandeer** it and be halfway to Port Royal before they even know it is gone.
There'd be a **warrant** out for our capture immediately!
The crew feels uneasy about Jack's plan, but a group of 13 pirates agrees to go with Jack.

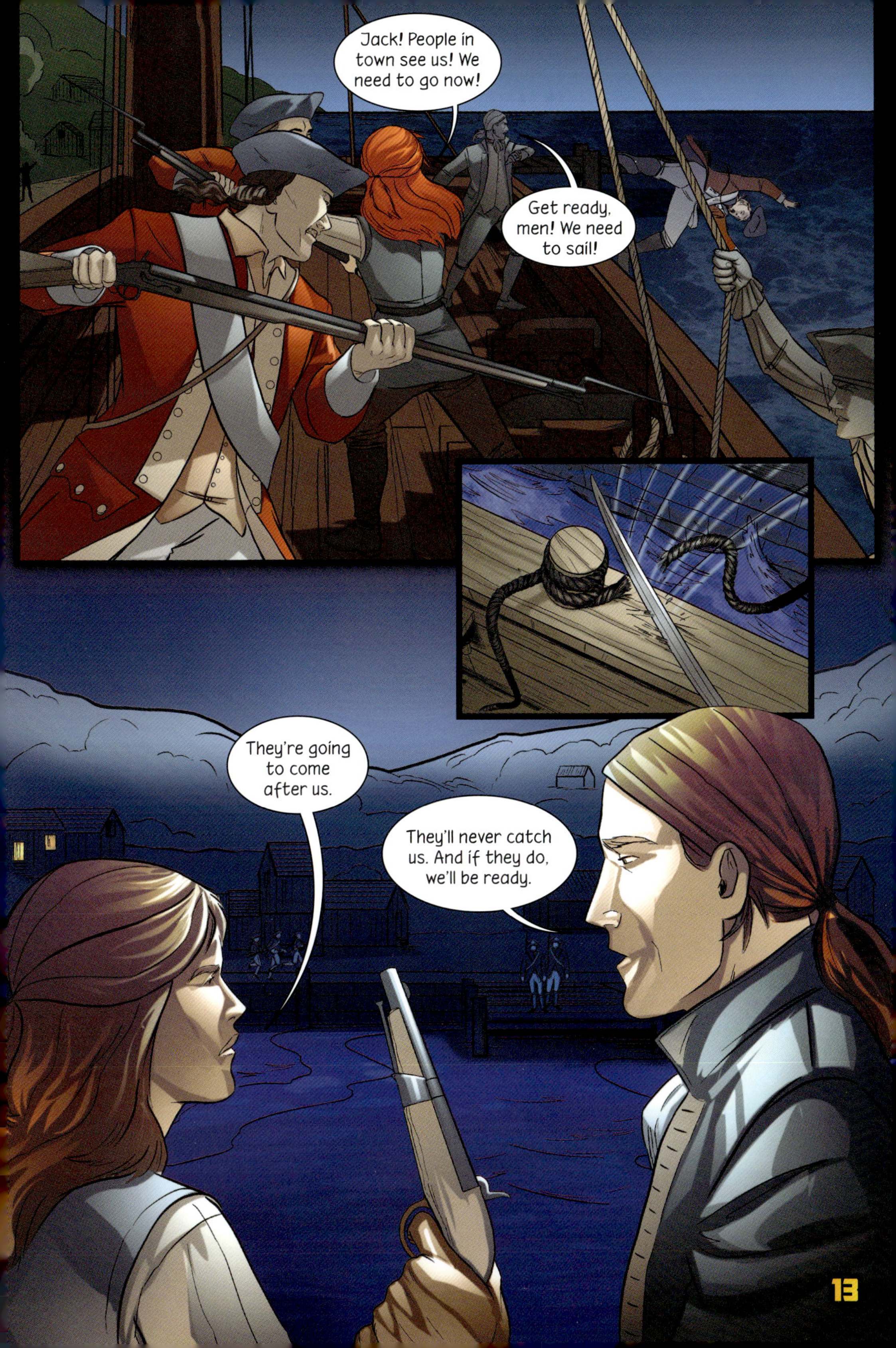
Jack! People in town see us! We need to go now!
Get ready, men! We need to sail!
They're going to come after us.
They'll never catch us. And if they do, we'll be ready.

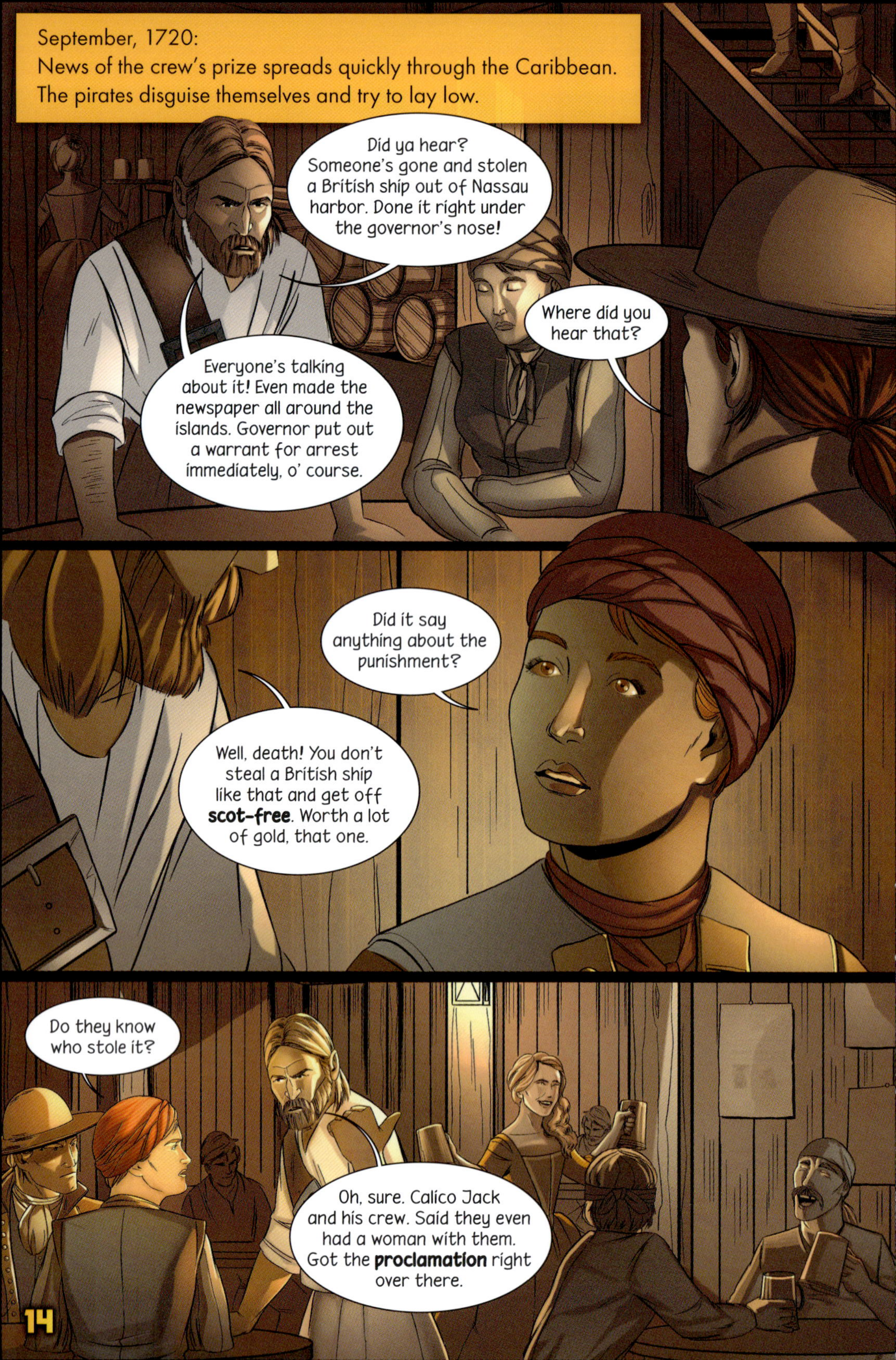
September, 1720:
News of the crew's prize spreads quickly through the Caribbean. The pirates disguise themselves and try to lay low.
Did ya hear? Someone's gone and stolen a British ship out of Nassau harbor. Done it right under the governor's nose!
Where did you hear that?
Everyone's talking about it! Even made the newspaper all around the islands. Governor put out a warrant for arrest immediately, o' course.
Did it say anything about the punishment?
Well, death! You don't steal a British ship like that and get off **scot-free**. Worth a lot of gold, that one.
Do they know who stole it?
Oh, sure. Calico Jack and his crew. Said they even had a woman with them. Got the **proclamation** right over there.

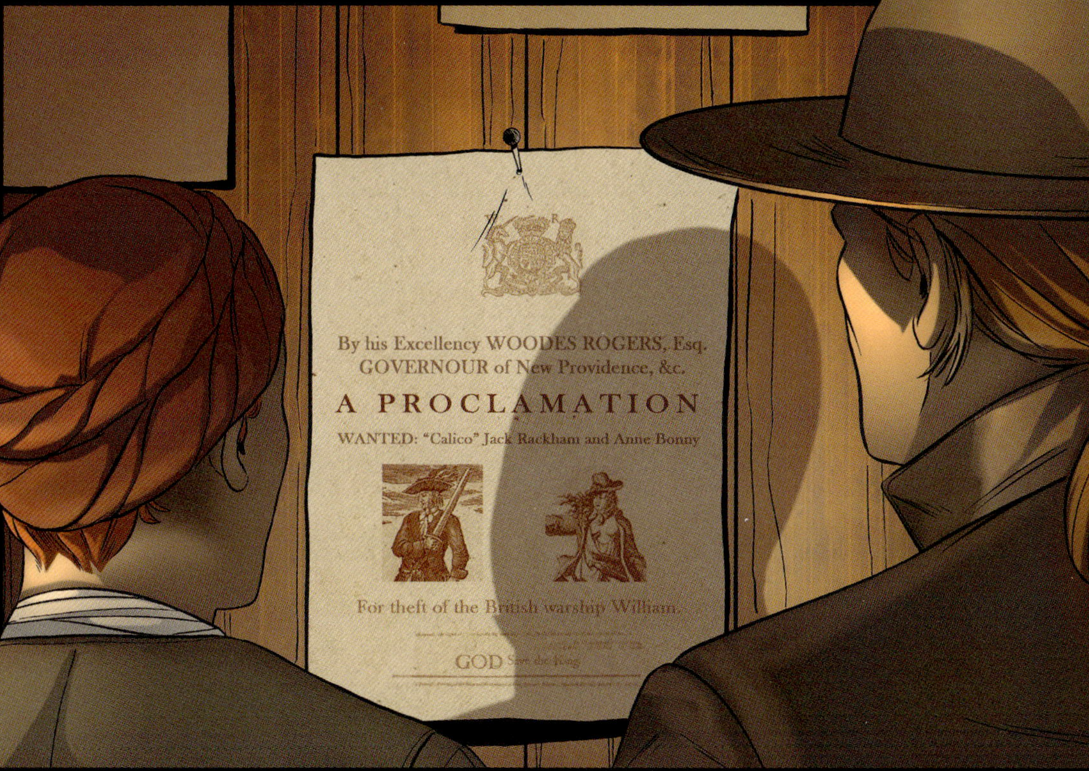
By his Excellency WOODES ROGERS, Esq.
GOVERNOUR of New Providence, &c.
A PROCLAMATION
WANTED: "Calico" Jack Rackham and Anne Bonny
For theft of the British warship William.
GOD Save the King

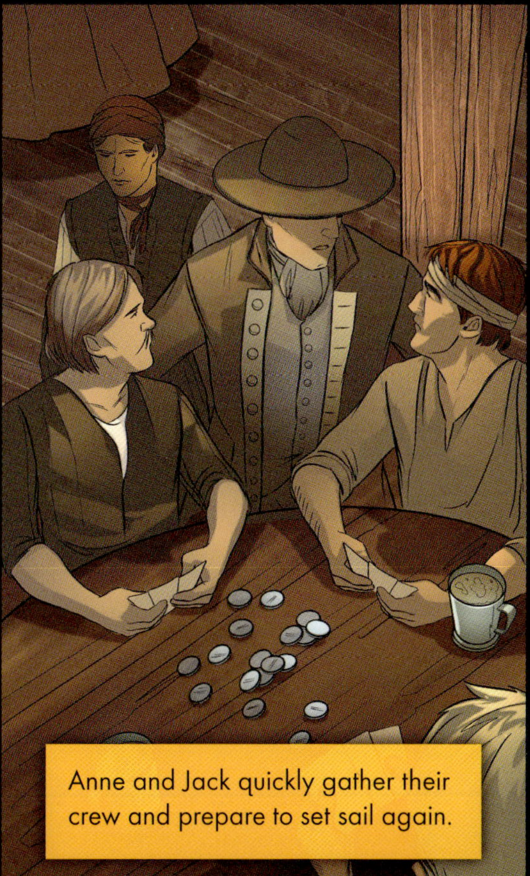
Anne and Jack quickly gather their crew and prepare to set sail again.

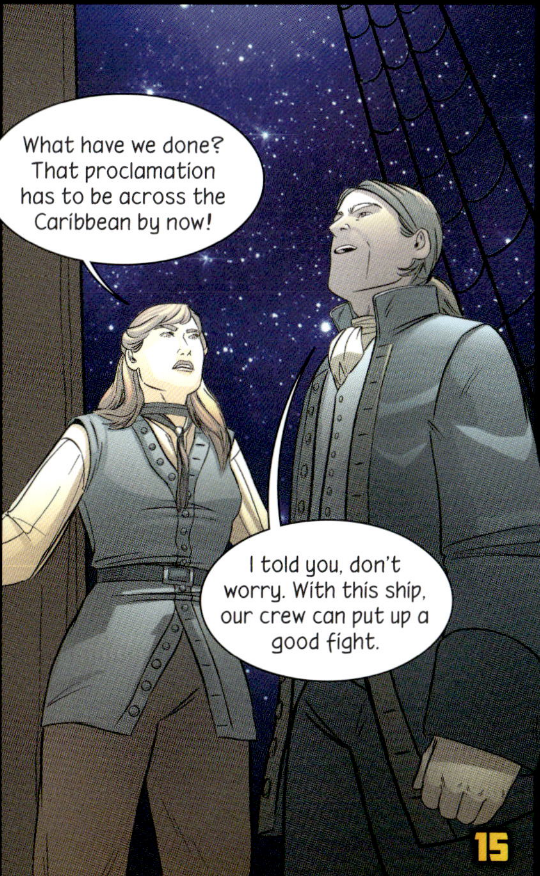
What have we done? That proclamation has to be across the Caribbean by now!
I told you, don't worry. With this ship, our crew can put up a good fight.

CAPTURED!

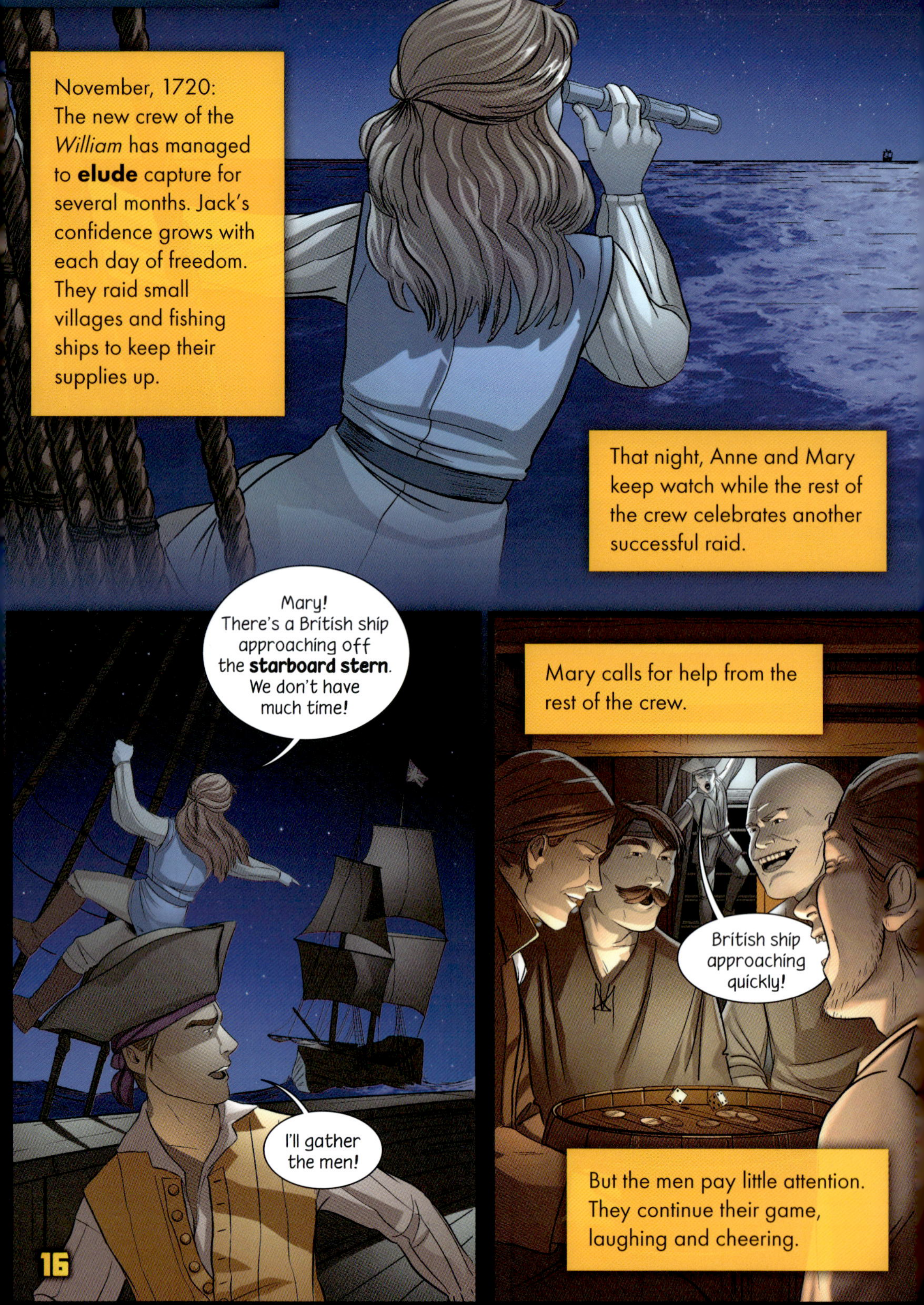
November, 1720: The new crew of the *William* has managed to **elude** capture for several months. Jack's confidence grows with each day of freedom. They raid small villages and fishing ships to keep their supplies up.
That night, Anne and Mary keep watch while the rest of the crew celebrates another successful raid.
Mary! There's a British ship approaching off the **starboard stern**. We don't have much time!
I'll gather the men!
Mary calls for help from the rest of the crew.
British ship approaching quickly!
But the men pay little attention. They continue their game, laughing and cheering.

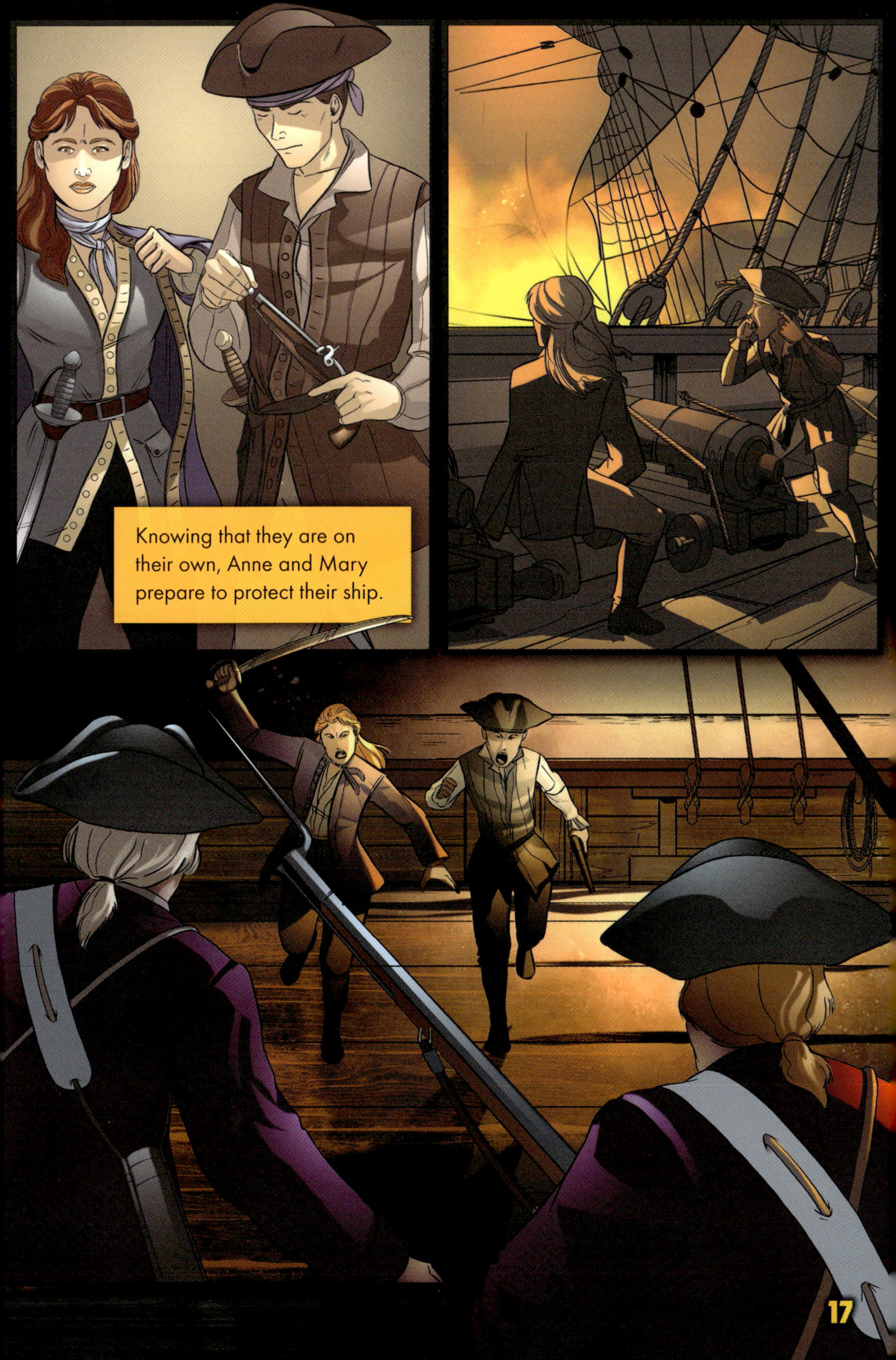
Knowing that they are on their own, Anne and Mary prepare to protect their ship.

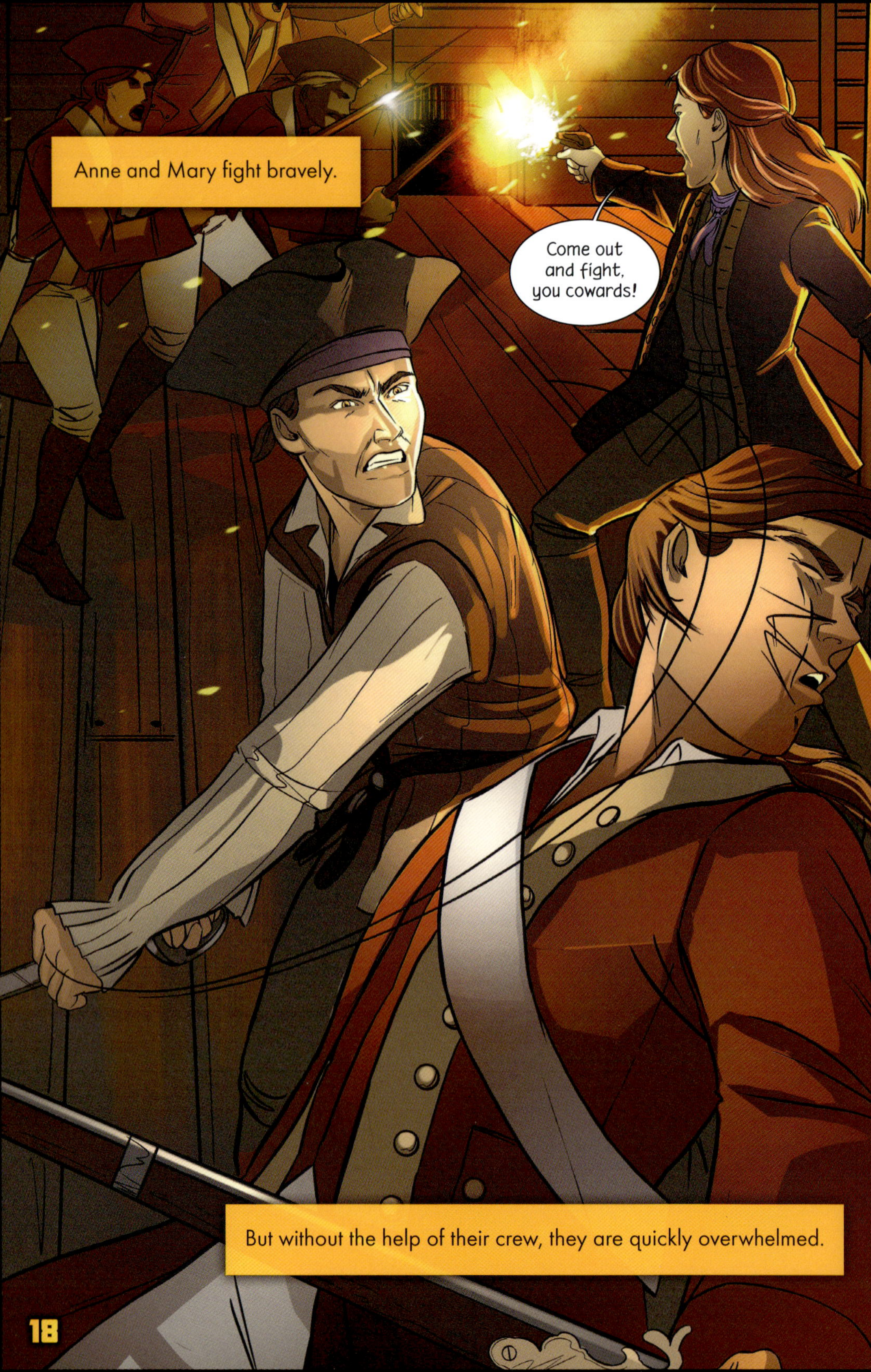
Anne and Mary fight bravely.
Come out and fight, you cowards!
But without the help of their crew, they are quickly overwhelmed.

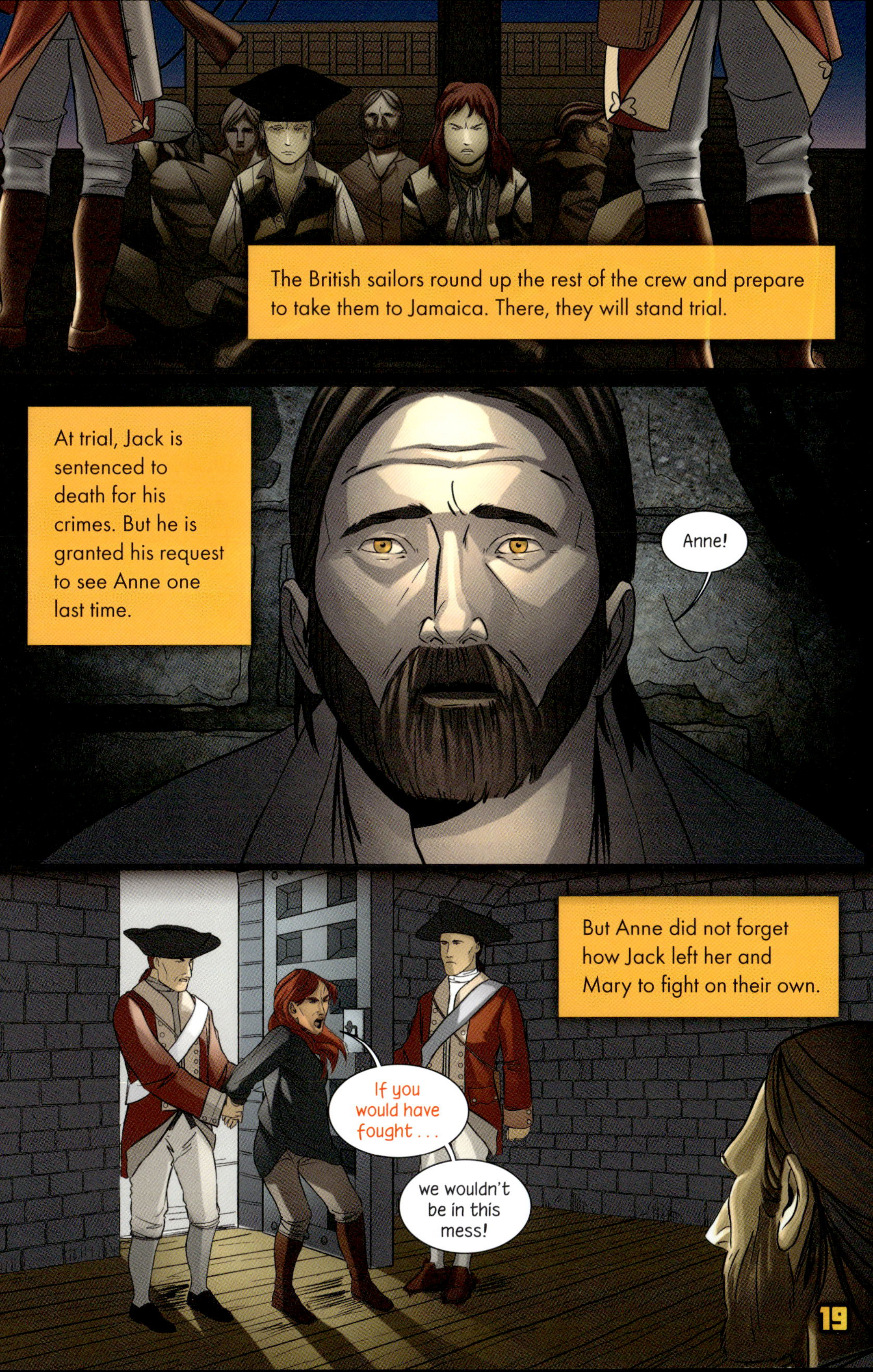
The British sailors round up the rest of the crew and prepare to take them to Jamaica. There, they will stand trial.
At trial, Jack is sentenced to death for his crimes. But he is granted his request to see Anne one last time.
Anne!
But Anne did not forget how Jack left her and Mary to fight on their own.
If you would have fought . . .
we wouldn't be in this mess!

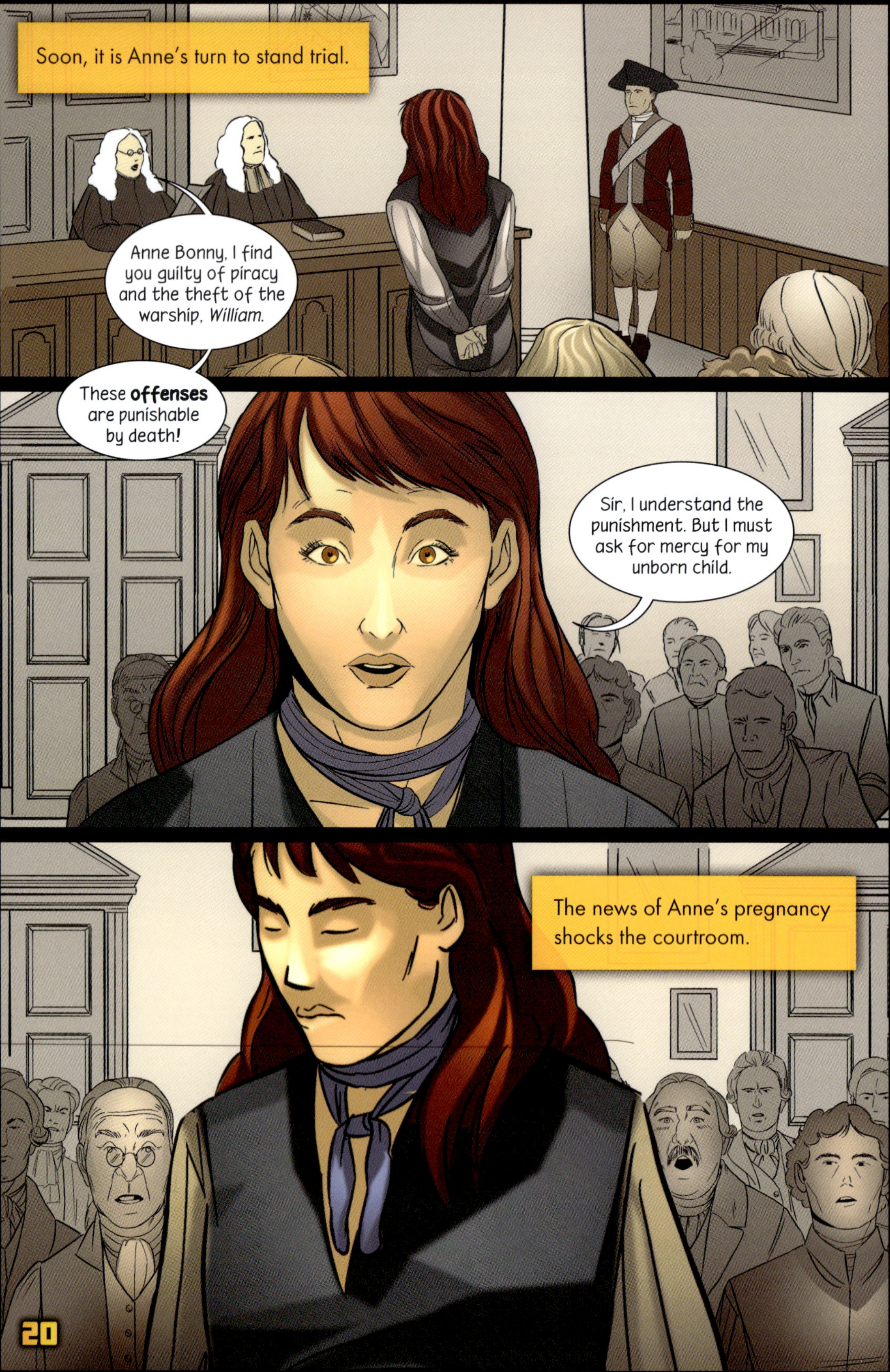

Soon, it is Anne's turn to stand trial.
Anne Bonny, I find you guilty of piracy and the theft of the warship, *William*.
These **offenses** are punishable by death!
Sir, I understand the punishment. But I must ask for mercy for my unborn child.
The news of Anne's pregnancy shocks the courtroom.

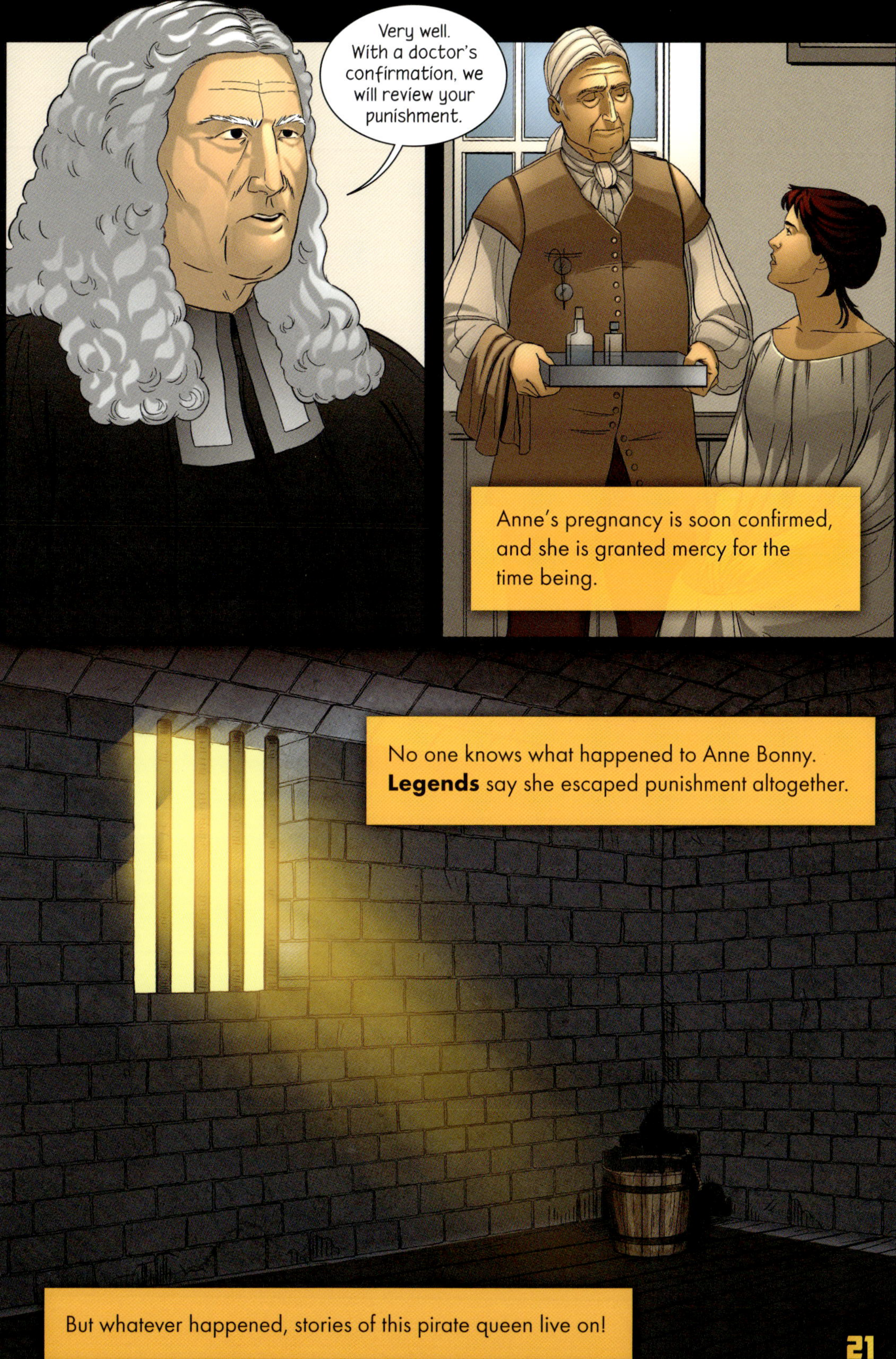
Very well. With a doctor's confirmation, we will review your punishment.
Anne's pregnancy is soon confirmed, and she is granted mercy for the time being.
No one knows what happened to Anne Bonny. **Legends** say she escaped punishment altogether.
But whatever happened, stories of this pirate queen live on!

More About Anne Bonny

- Much mystery hangs around what happened to Anne. Most people believe her father paid for her release. Some sources say she remarried and had eight children.
- The main source of information about Anne Bonny's life comes from *A General History of the Robberies and Murders of the Most Notorious Pyrates*. Many people believe it was written by Daniel Defoe, who wrote the fictional *Robinson Crusoe*. No one knows how much of the book is true.

Timeline

Between 1719 and 1720: Anne joins Calico Jack's crew

August 1720: Anne helps Calico Jack commandeer a British sloop

November 15, 1720: Anne and the crew are captured by pirate hunters

November 28, 1720: Anne goes on trial for piracy

Anne Bonny's Travels

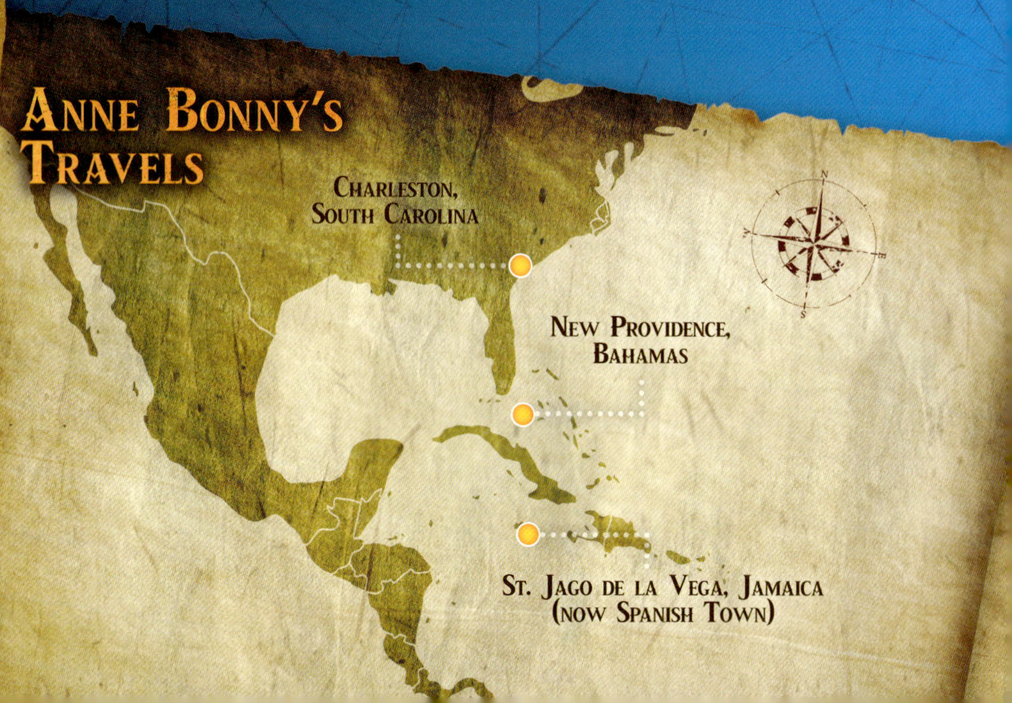
Charleston, South Carolina
New Providence, Bahamas
St. Jago de la Vega, Jamaica (now Spanish Town)

Glossary

commandeer—to steal or take by force

elude—to escape notice

fore rigging—ropes and lines toward the front of a ship that support sails and masts

legends—stories that are not able to be proven as true

letter of marque—a document that gives ships permission to capture enemy merchant ships

offenses—acts that are against the law

plunder—to take goods by force

privateer—a sailor on a ship who is licensed to attack enemy ships

proclamation—a formal public announcement

raids—surprise attacks that use force

ransack—to forcefully search through and steal from

scot-free—free from harm or responsibility

sloop—a small ship that has one mast

starboard stern—the back right of a ship

warrant—a document that gives permission to authorities

To Learn More

At the Library

Leaf, Christina. *Mary Read: Pirate in Disguise*. Minneapolis, Minn.: Bellwether Media, 2021.

Perdew, Laura. *Anne Bonny*. Hockessin, Del.: Mitchell Lane Publishers, 2016.

Steer, Dugald A. *Pirateology*. Cambridge, Mass.: Candlewick Press, 2006.

On the Web

Factsurfer.com gives you a safe, fun way to find more information.

1. Go to www.factsurfer.com
2. Enter "Anne Bonny" into the search box and click 🔍.
3. Select your book cover to see a list of related content.

Index

Bonny, John, 4
Caribbean, 14, 15
Charleston, South Carolina, 4
historical quotes, 19
Ireland, 4
Jamaica, 19
New Providence, Bahamas, 4, 12
Port Royal, 12
privateer, 4, 6, 12
proclamation, 14, 15

Rackham, John (Calico Jack), 4, 5, 6, 8, 9, 10, 11, 12, 13, 14, 15, 16, 19
Read, Mary, 8, 9, 10, 11, 12, 16, 17, 18, 19
Revenge, 5, 7, 8, 12
timeline, 22
travels, 22
trial, 19, 20
warrant, 12, 14
William, 12, 16, 20